YOUR KNOWLEDGE HAS VALUE

AF151760

- We will publish your bachelor's and master's thesis, essays and papers

- Your own eBook and book - sold worldwide in all relevant shops

- Earn money with each sale

Upload your text at www.GRIN.com and publish for free

Matthias Zöphel, Christian Egger, Hansjakob Riedi

A Short Critical, Non-Technical, Non-Mathematical Paper about Regression Analysis

An Introduction for Beginners

GRIN Publishing

Bibliographic information published by the German National Library:

The German National Library lists this publication in the National Bibliography; detailed bibliographic data are available on the Internet at http://dnb.dnb.de .

This book is copyright material and must not be copied, reproduced, transferred, distributed, leased, licensed or publicly performed or used in any way except as specifically permitted in writing by the publishers, as allowed under the terms and conditions under which it was purchased or as strictly permitted by applicable copyright law. Any unauthorized distribution or use of this text may be a direct infringement of the author s and publisher s rights and those responsible may be liable in law accordingly.

Imprint:

Copyright © 2008 GRIN Verlag, Open Publishing GmbH
Print and binding: Books on Demand GmbH, Norderstedt Germany
ISBN: 978-3-640-90961-2

This book at GRIN:

http://www.grin.com/en/e-book/171083/a-short-critical-non-technical-non-mathe-matical-paper-about-regression

GRIN - Your knowledge has value

Since its foundation in 1998, GRIN has specialized in publishing academic texts by students, college teachers and other academics as e-book and printed book. The website www.grin.com is an ideal platform for presenting term papers, final papers, scientific essays, dissertations and specialist books.

Visit us on the internet:

http://www.grin.com/

http://www.facebook.com/grincom

http://www.twitter.com/grin_com

HTW Chur
Hochschule für Technik und Wirtschaft
University of Applied Sciences

Self Study Assignement in „Quantitative Methods"

Master of Science in Business Administration

A Short Critical, Non-Technical, Non-Mathematical Paper

about

Regression Analysis

Group:

Matthias Zöphel
Christian Egger
Hansjakob Riedi

Index

INTRODUCTION

The following report will provide an insight into regression analysis based on three sections. First, the technique will be described in a non-mathematical way by indicating a six-stages-procedure which is used in Hair's multivariate analysis 3rd edition. For understanding and extending reasons several other sources have been incorporated into this assignment. The second section will identify limitations to regression analysis indicating when it is appropriate to use and what limitations arise once it is used. Finally, the third section of this report will provide two research examples which are established according to the six-stage-procedure exemplified in the technique description section.

1. REGRESSION ANALYSIS - TECHNIQUE DESCRIPTION

Regression Analysis is one technique component of multivariate analysis and is used when the relationship between several predicting or independent variables and a dependent or criterion variable is to be analyzed (Pearson 1908). In contrast to simple regression which incorporates only a single independent variable, multiple regressions incorporate several independent variables with the advantage of achieving a higher level of prediction of the dependent variable. The second intended outcome of multiple regressions besides predicting the dependent variable is to identify correlations between independent variables thus explaining which independent variable has what contribution in predicting the dependent variable. The purpose of these two intentions of regression analysis is to enable organizations to create knowledge and thereby improve decision-making (Hair 2006). The regression analysis technique incorporates six stages which are indicated below, leading to the prediction of the predicting variable.

Stage 1: Stating the Research Problem

The first stage in multiple regression analysis is to determine the researcher's objective meaning that a dependent variable must be selected which the researcher wishes to have predicted. Once the researcher knows what he wants to predict, he selects variables that he considers to be influential to the dependent variable, variables that have explanatory intention of why changes in these independent variables influence the dependent or predicting variable. Based on the researchers theoretical knowledge he can assume which of the independent variables he has selected carries what weight in predicting the dependent variable. Essential to use multiple regression analysis is the selection of metric or quantitative variables and no non-metric or qualitative variables unless they are transformed to dummy variables which this report will refer to later on. When selecting the variables, the researcher must be aware of specification errors which are either the inclusion of a non-essential variables or the omission of an essential one which in both cases can lead to a non-model parsimony which further leads to a fraud outcome of the regression analysis. Once the researcher knows what he would like to have predicted and by what means, he has to collect data being the second step of regression analysis.

Stage 2: Designing the Research

The size of data needed depends on the number of independent variables the researcher intends to incorporate into his regression model. It is suggested (Hair 2006), that for each independent variable, the sample size should not fall below a ratio of 5:1. However, the desired ratio is 15 to 20:1 meaning that there have to be 15-20 data collections available for each independent variable. The higher the ratio, the higher will be the degree of freedom and statistical power which are in charge of achieving generalizability of the results. As degree of freedom and statistical power reflect generalizability, the researcher can raise the degree by either excluding independent variables or increasing the sample size which both ways lead to a greater ratio and thus to a greater degree of freedom and statistical power and thus to more generalizability. However, the researcher must also be careful when selecting a sample size that is too great as this can also make the regression analysis overly sensitive. Data can be collected through surveys, questionnaires or other means. The researcher must pay attention to use questions that are appropriate and lead to the most reliable outcome. If this is not done appropriately, or if data once collected is not appropriately used, or if some data is missing then this is referred to a measurement error leading to invalid and unreliable results. However, it is not possible to eliminate the measurement error as there will always be some problem when designing the research. Still the researcher is required to minimize measurement error and thus trying to reach out for the most acceptable level of validity. Once the researcher has collected the necessary data, he inserts them into statistical software which helps him to identify whether his data meets the assumptions required by regression analysis stated now in stage three.

Stage 3: Meeting the Assumptions

The four assumptions in multivariate analysis are linearity, homoscedasticity, Normality and Independence of the error terms. A violation of these assumptions will lead to invalid outcomes. Therefore, the researcher is required to test the assumptions even twice: first for each separate variable and second for the overall model (Hair 2006). When one of the assumptions is not met, the researcher needs to undertake data transformation.

i. *Normality*

Normality is the most fundamental assumption of regression analysis referring to the shape of the data distribution for an individual metric variable and its correspondence to the normal distribution (Hair 2006). Normality can be identified by the help of statistical tests, histograms or normal probability plots which last one is rather used for smaller sample sizes. Smaller sample sizes are even the main reason why non-normality can occur. Histograms are usually used for identification and when non-normality exists then this can be seen by a flat of peaked curve or by an unbalanced curve that is either shifted to the right or to the left. Outliers are often the reason for a non-normal distribution. Outliers occur when a data is incorporated into the analysis that deviates tremendously from the other collected data. When a non-normal distribution was identified by the diagrams, the researcher is required to undertake remedies to achieve normal patterns of distribution again. Taking the inverse, square-root, logarithms, least- or cubed squares are the transforming remedies that can be applied. For deciding which of the stated remedies to use depends on the non-normal pattern of the curve. However, it is often a trial and error process the researcher has to undergo. In case of outliers, it is even often preferable to delete that data from the analysis in order to achieve normal distribution.

     ii.   _Homoscedasticity_

Homoscedasticity refers to the assumption that the dependent variable exhibits equal levels of variance across the range of predictor variables. Thus, if homoscedasticity is not meet then this will lead to an unfair testing of the relationship across all values of the non-metric variables (Foster 2006). In order to identify whether homoscedasticity exists, the researcher is best advised to use scatterplots. When most of the values are in the middle range of the scatterplot or when one or more variables seem to be skewed, the researcher is confronted with heteroscedasticity and is required to undertake remedies in order to achieve homoscedasticity again. Taking the inverse or the square root are also the actions required to be taken here. When the scatterplot indicates a cone opened to the right, the researcher must take the inverse, when it is opened to the left, he must take the square root.

     iii.   _Linearity_

Linearity means that changes in the independent variable must be proportionate to changes in the dependent variable. If this is not the case it is called to be curvilinear leading to an invalid outcome of the results. Linearity can best be identified by scatterplots and is met when the plots indicate a linear line. When curvilinearity exist, the researcher is best off by creating polynomials. Polynomials are power transformations of an independent variable that add a non-linear component for each additional power of the independent variable (Stamatis 2003). Thus, through adding powers to the variables, the researcher is able to meet the assumption of linearity.

iv.   *Interdependence of the error terms*

As the term already clarifies, interdependence of the error terms is the absence of any correlation between prediction errors. When scatterplot patterns indicate all errors to be positive while on the other hand the alternative values are negative, the researcher can be certain to have identified correlation between error terms (Gilmartin and Hartka 1992). Interdependence of error terms is related to multicollinearity in which independent variables are correlated with each other. When errors occur while multicollinearity exists then this makes it difficult for the researcher to identify which independent variable causes what contribution to the error, thus making him either underestimate or overestimate the effect that one independent variable has on the dependent variable. In order to remedy a violation of this assumption, the researcher is often advised to add a moderator effect that indicates the interdependence. A moderator effect is most of the time a dummy variable stating that the interdependence is either caused by a) or b). This can be for example a gender case indicating that males (a) have a different effect on the outcome than women (b).

Once each variable and the overall variate have been tested for assumptions and remedies have been successfully implied, the researcher proceeds to stage four in which he estimates the regression model.

Stage 4: Estimating and Assessing Model

The researcher has measured all variables in the first three stages and will now decide what independent variables to include in the variate with the intention to achieve highest statistical and practical significance of the prediction. In order to do so, the researcher can either himself select the variables through the confirmatory approach based on the data results from the previous steps or he utilizes a regression procedure through sequential search methods or through the combinatorial approach which both choose the independent variables for him based on best regression model. The three approaches that intent to find the best regression model are now discussed below.

i.   _Confirmatory Specification_

The researcher decides himself which independent variables to include in the regression model based on his theoretical knowledge he gained in the first three steps. While selecting the independent variables to be included, the researcher must greatly pay attention to the specification error that leads to disharmony of the model.

ii.   _Sequential Search Method_

In this approach, variables are either included one after the other starting with the one that has the highest correlation to the dependent variable or by including all at once and then excluding one after the other. Always when one variable is either included or excluded, the researcher is supposed to look at the statistical significance level of the variate in order to see whether it has increased or decreased and based on this to decide what independent variables are relevant or irrelevant in the variate. A big hurdle here is the effect of multicollinearity which happens when a single independent variable on its own does not raise the significance level but does so in combination with another independent variable.

iii.   _Combinatorial Approach_

This approach identifies the model that has the greatest statistical significance through a combination of all possible inclusions. This is easily done with statistical programs but has been criticized as it is even greater neglecting the effects of multicollinearity and also of outliers.

Once the variables are chosen to be included in the regression model, the researcher tests them for the assumptions in stage three again. When he has the confirmation that assumptions are met, he tests the significance of the four regression coefficients which will be to ensure that results can certainly be generalized to other populations. The four coefficients are the coefficient of determination, the adjusted coefficient of determination, the standard error of the estimate and the statistical significance of the regression coefficients.

Stage 5: Interpreting the Regression Variate

The researcher must evaluate the estimated regression coefficients for their explanation of the dependent variable (Hair 2006). This evaluation serves two purposes. First of all the researcher is now able to estimate and forecast the dependent variable by using the coefficients and secondly and even more important, the researcher will be able to explain which independent variable contributes to what extend to the prediction of the dependent variable which both purposes are the reasons for regression analysis. The second purpose is made possible by the beta coefficients which are the regression coefficients in a standardized form. When the beta coefficients are standardized to a common scale then variables become comparable and thus become explainable towards what extend they contribute towards predicting the independent variable.

Stage 6: Validating the Results

Validating the results is the last stage of the regression analysis in which is ensured that the regression model is once more estimated for generalizability by using a different population and then comparing it with the one under investigation.

2. USAGE AND LIMITATIONS OF REGRESSION ANALYSIS

As already mentioned, regression analysis is the most widely used dependence technique as it is applicable in every facet of business decision making. Its uses range from the most general problems to the most specific, in each instance relating factors to a specific outcome (Hair 2006). However, there are still certain limitations to multiple regression analysis that can prevent the use of regression analysis or distort the outcome of prediction when using it. We divided limitations up into two categories. The first category states limitations that prevent the use of regression analysis while the second category represents limitations that arise while regression analysis is used mainly referring to the assumptions of regression analysis.

a) Limitations preventing the use of Regression Analysis

Regression analysis is one component of multivariate analysis and can only be used when several preconditions are met. The first precondition demanded is that the analysis will measure a dependent variable making regression analysis a dependence technique that cannot be used for measuring interdependence relationships. Even though when a dependent relationship is measured, it still limits regression analysis to one dependent variable to be predicted only. The analysis is thus not capable of measuring several dependent variables simultaneously indicating the second precondition that must be met. Lastly, regression analysis is only capable of measuring metric data making any qualitative data usages not possible. Once these preconditions are met, regression analysis can be used but still incorporates limitations within its usage which is referred to below.

b) Limitations arising while using Regression Analysis

i) *Causality / Multicollinearity*

The major conceptual limitation of all regression techniques is that one can only ascertain relationships, but never be sure about underlying causal mechanism (Ruspini 2002). What Ruspini means by this is that when a dependent variable indicates great association with the independent variable then this still does not necessarily mean that a change in the independent variable results in a change in the dependent variable. This is mainly due to multicollinearity which means that independent variables are correlated with each other and mutually result in a change of the dependent variable. Regression analysis does not provide valid outcomes under such circumstances.

ii) *Sample Size*

As already stated in the technique description, the sample size for the regression analysis must be large enough to ensure generalizability. When sample sizes fall below a ratio of 5:1 towards the independent variables then outcomes will not be valid. On the other hand, sample sizes that are too large can also make the model overly sensitive also disturbing a reliable outcome of the results. It is therefore necessary to bear in mind that statistical significance which is achieved by large sample sizes, does not necessarily imply practical significance (Foster, Barkus et al. 2006.). This regression analysis does only provide valid outcomes with the right balance of the sample size.

iii) *Outliers / Linearity*

Outliers are defined as values that are at least 3 standard deviations above or below the mean (Sanderson 2001). Regression Analysis is only able to describe linear relationships between dependent and independent variables. However, the analysis is not resistant to outliers. If the researcher does not filter his data to remove outliers first, the line will be shifted toward the outlier and will reduce the squared error thus making the data unreliable (Wisnowski 1999). Outliers need either to be transformed or deleted from the analysis to make it provide a reliable outcome.

iv) *Measurement / Specification Error*

When multicollinearity or outliers are present, a solution would be to drop them from the analysis. However, sometimes the dropping of a variable leads to more severe effects on result's validity which is referred to the specification error. Specification error also refers to the inclusion of a variable that is irrelevant and can prevent a reliable outcome. In both cases, the result will be that ordinary least squares estimators will no longer be unbiased, and the bias will depend directly on the size of the correlation between the errors and exogenous variables (Kaplan 1955). Regression analysis will thus be limited in its usage when specification errors are present and not remedied.

Besides the specification error there is also the measurement error that always occurs in regression analysis as there will always be some problems with data collection. As a result, measurement error tends to attenuate the variable's regression coefficients and tends to make the variable an imperfect statistical control (Fox 1997). When measurement errors are not minimized, results will be of little generalizability. As measurement errors cannot totally be controlled, it stands for a major limitation of regression analysis.

3. RESEARCH EXAMPLES

3.1. The Effect of a Quality Management System on Supply Chain Performance: An Empirical Study in Taiwan" (Liu 2009)

Introduction

This Research examines, whether the attitude of firms in Taiwan concerning the implementation of ISO/TS 16949 quality management system has really a positive effect on the supply chain performance. For this reason, three Hypothesis were stated, but only Hypothesis two deals with the multiple regression analysis and will therefore be used as research example in this report. This research example will be exemplified by using the six-stage-procedure from the technique description section.

a) *Stating the Research Problem: Stage 1*

Hypothesis 2: *Implementation of the quality management system ISO/TS 16949 has a positive effect on supply chain performance in the company and relations of the company with its suppliers and customers.*

b) *Designing the Research: Stage 2*

The data for the analysis of all three hypotheses were collected by the same questionnaire based on the SCOR model and their criterion's which measures the performance of the supply chain. The answers were given using a Likert five-point scale from strongly disagree (1) to strongly agree (5) what means, that they made nonmetric variables metric by transforming them to ordinal scales and therefore rank able as mentioned above.

c) *Meeting the Assumptions: Stage 3*

The research paper examined did skip this stage and did not try to meet the necessary assumptions. According to Osborne and Waters (2002), few articles report having tested assumptions of the statistical tests they rely on for drawing their conclusions. Therefore, we have to call into question the validity of many of these results, conclusions, and assertions, as we have no idea whether the assumptions of the statistical tests were met. This statement can be referred to this research paper, thus there is no guarantee for valid results in this research example.

d) *Estimating and Assessing Model: Stage 4*

The research paper chose the confirmatory approach for estimating the research model.

To test hypothesis 1, a factor analysis was performed. From the result of the factor analysis was four factors extracted. These four supply chain performance factors, which were used as independent variables are named as: X1) Expenses of cost, X2) Assets/utilization, X3) Supply chain reliability and X4) Flexibility and responsiveness. On the other hand, the managerial satisfaction with their supply chain after implementing the QMS served as the dependent variable.

e) *Interpreting the Regression Variate*

The result of the analysis demonstrate, that three out of four supply chain performance factors are associated with the managerial satisfaction. These three independent variables are: X2) Assets/utilization, X3) Supply chain reliability and X4) Flexibility and responsiveness. These three had a significant positive relation (alpha = 5% or 0.05 / r = 0.627) to the managerial satisfaction after having implemented the QMS according to ISO/TS 16949. The factor with the most significant effect on the dependent variable was the independent variable X3) Supply chain reliability.

f) Validating the Results

As well as with the assumptions in stage three, we could also not see any indication that results validation has been performed. Thus, another step was skipped by the researcher making the results even more questionable of reliability.

Conclusion

This research demonstrates the usage of multiple regression analysis by using 4 independent variables which had to be transformed from nonmetric to metric variables, resulting in three out of four independent variables with a positive relation to the dependent variable and therefore proofing the hypothesis 2. However, as two of the six steps have been skipped by the researcher among which is the important one of meeting the assumptions; we cannot consider this research as valid although we might be able to assume that it has been undertaken implicitly due to its publishing in the journal of management.

3.2 Applying the Theory of Planned Behavior (TPB) to Predict Internet Tax Filing Intentions (Ramayah, Yusliza et. al 2009)

Introduction:

The objective of this Research is to model the intention of tax payers in Malaysia to file a tax return either manually or by email applying the TPB. This research paper again will be analyzed according to the six-stage-procedure stated in the technique description in section two of this report.

a) Stating the Research Problem: Stage 1
"In psychology, the TPB is a theory about the link between attitudes and behavior. It was proposed by leek Aj zen (1991). The TPB is an extension of Theory of Reasoned Action (TRA) made necessary by the original model's limitations in dealing with behaviors over which people have incomplete volitional control (Ramahja 2009)."

Stage 1 was performed by defining objective of this Research, which is to model the intention of tax payers in Malaysia to file a tax return either manually or by email applying the TPB. You may consider the following statement as explanation for the TPB model: "TPB holds that human action is guided by three kinds of considerations which are behavioural beliefs, normative beliefs and control beliefs. Based on this model, the following three hypotheses were predicted:

Hypothesis 1: *Attitude toward using e-filing to pay tax is positively related to intention to use e-filing.*

Hypothesis 2: *PBC is positively related to intention to use e-filing.*

Hypothesis 3: *SN is positively related to intention to use e-Filing.*

You may consider the following statement as explanation for the term "SN": *SN is a function of beliefs about the expectations of important referent others, and his/her motivation of complying with these referents. SN also determined by beliefs that specific other persons advocate performing or not performing the behavior; and the motivation to comply with their wishes.*

b) *Designing the Research: Stage 2*

The Research design was undertaken by a sample of 125 tax-paying employees from different regions within Malaysia were all of them need to complete a questionnaire specially developed for this Research study.

c) *Meeting the Assumptions: Stage 3*

The Researchers rely on prior studies for each hypothesis. For hypothesis 1, the Researchers argue that "prior studies have shown that attitude does have a significant and positive impact on intention" (from Research Example 2). They refer to an example, where data from 239 Malaysian bank customers were analyzed by a multiple regression regarding their attitude in using Internet banking. The findings were positively related to use Internet banking, what lead to the hypothesis 1. For the second hypothesis, the results of former Research showed "a positive relationship between PBC (Perceived Behavioural Control) and intention in various research contexts" (from Research Example 2) which leads to hypothesis 2. Finally, "mixed results have been obtained from previous empirical studies showing no significant relationship between SN and intention, while some showing significant relationship between subjective norm and intention" (from Research Example 2) which lead the Researcher to assume correctness of hypothesis 3. Instead of using variables as indicators like in Research example 1, the decision was made to use what they called short scale and we interpret as a variate. The independent variables in the model, X1) attitude toward using email for tax returns, X2) subjective control and X3) behaviour control, was measured by separate scales relied to existing scales from the literature. All variables were reviewed in order to achieve the highest possible validity and reliability by applying a factor analysis for validity and the alpha coefficient to test the Reliability. The results confirmed that no multicollinearity is existent. Due to former research that has been undertaken, we can assume that assumptions have been met in these studies. However, this study involves a far to different model to just assume the assumption meeting. We argue that assumptions should have been tested again for this research paper to ensure validity.

d) _Estimating and Assessing Model: Stage 4_
The Researchers focus on the behavioural intention of tax payers as the dependent variable and X1) attitude, X2) subjective norm and X3) perceived behavioural control as the independent variables.

e) _Interpreting the Regression Variate: Stage 5_
The results showed that the TBP is applicable in the Malaysian setting to significantly explain intention to use e-Filing

f) _Validating the Results: Stage 6_
As well as with the Research example 1, we could also not see any indication that results validation has been performed. Thus, another step was skipped by the researcher making the results even more questionable of reliability.

SUMMARY

In conclusion, regression analysis is one branch of multivariate analysis. Even though there are some preconditions that need to be met in order to use the method, it is still the most widely used technique among all multivariate techniques and is there to help organizations or individuals in the decision-making process. The technique consists of six stages which each of them must be undertaken appropriately in order to achieve met valid and generalizable results. Regression analysis is limited to its use when assumptions are not met or when measurement or specification errors occur. However, certain remedies can be applied that still lead the researcher to achieve the intended outcome. As exemplified by one of the research examples used, certain stages have not been followed with the result of unreliable outcome while the other research question has stuck to the six-stage-procedure giving us greater certainty of valid and generalizable results.

LITERATURE

Hair, J. Black, W. Babin, B. & Anderson, R., 2006. Multivariate Data Analysis: a global perspective. 7th ed. Pearson: Prentice Hall.

Sykes, A., 1999. An Introduction to Regression Analysis. Chicago Law and Economics Paper, (online). http://www.law.uchicago.edu/files/files/20.Sykes_.Regression.pdf (accessed 28 October 2009)

Ruspini, E., 2002. Introduction to Longitudinal Research. (e-book) Routledge: London. http://books.google.de/books?id=wcZ6oTxQoeAC&pg=RA1-PA11&lpg=RA1-PA11&dq=For+example,+a+real+estate+agent+might+record+for+each+listing+the+size+of+the+hou se&source=bl&ots=l5agu_p3wM&sig=sbRDVr--ZLArXI8NL8clVbFS5wE&hl=de&ei=ctnhSqq0B5Te-QbGlazoAQ&sa=X&oi=book_result&ct=result&resnum=3&ved=0CBAQ6AEwAg#v=onepage&q=F or%20example%2C%20a%20real%20estate%20agent%20might%20record%20for%20each%20listin g%20the%20size%20of%20the%20house&f=false (Accessed 26 October 2009).

Gilmartin, K. and Hartka, E., 1992. Using Regression Analysis to compute Pay Back. Journal of Law, Science and Technology, (online). 32 (3). http://www.air.org/publications/documents/backpay.pdf (Accessed October 30 2009).

Fox, J., 1997. Applied regression analysis, linear models, and related methods. Sage Publications: Thousand Oaks.

Kaplan, D., 2006. Structural equation modeling: foundations and extensions. Sage Publications: Thousand Oaks.

Foster, B. et al., 2006. Understanding and Using Advanced Statistics. Sage Publications: Thousand Oaks.

Stamatis, D., 2003. Six Sigma and Beyond: Statistics and Probability. St. Lucie Press: Florida.

Osborne, J. and Waters, E., 2002. Four assumptions of multiple regression that researchers should always test. Practical Assessment Research and Evaluation Paper, (online). 8(2). http://pareonline.net/getvn.asp?v=8&n=2 (Accessed November 3 2009).

Osborne, Jason & Elaine Waters (2002). Four assumptions of multiple regression that researchers should always test. Practical Assessment, Research & Evaluation, 8(2). Retrieved November 6, 2009 from http://PAREonline.net/getvn.asp?v=8&n=2 . This paper has been viewed 95,492 times since 1/7/2002.

Sanderson, M., 2001. Example pof Problem Involving Outliers. http://www.herkimershideaway.org/writings/n1ap01.htm (Accessed November 4 2009)

Sources for Research Examples:

Liu, C., 2009. The Effect of a Quality Management System on Supply Chain Performance: An Empirical Study in Taiwan. International Journal of Management. 26(2). http://proquest.umi.com.ezproxy.fh-htwchur.ch/pqdweb?index=0&did=1874986631&SrchMode=1&sid=2&Fmt=3&VInst=PROD&VTyp e=PQD&RQT=309&VName=PQD&TS=1257529136&clientId=58690 (Accessed November 3 2009)

Ramayah, T., 2009. Applying the Theory of Planned Behavior (TPB) to Predict Internet Tax Filing Intentions. International Journal of Management. 26(2). http://proquest.umi.com.ezproxy.fh-htwchur.ch/pqdweb?index=5&did=1874986621&SrchMode=1&sid=2&Fmt=3&VInst=PROD&VTyp e=PQD&RQT=309&VName=PQD&TS=1257457231&clientId=58690 (Accessed November 3 2009).